# The Aesthetic Sense and Religious Sentiment in Animals

# The Aesthetic Sense and Religious Sentiment in Animals

Louis Viardot
Edward P. Evans
Charles G. Allen

# Part I

## On the Aesthetic Sense in Animals[1]

The mind of animals is a very old subject of discussion. Descartes and his school regarded an animal as a mere piece of machinery, like a clock or a turnspit. For man alone they reserved intelligence, meaning by that, memory, feeling, will, and reason. The story of Malebranche is well known: As he was going into his convent at the Oratory with a friend, a little bitch ran up and fawned on him; he gave her a kick which sent the poor beast yelping off, and when his friend expressed surprise that so gentle, kindly, and Christian a person returned kicks for caresses, he exclaimed,

[1] By Louis Viardot. *Translated by A. R. Macdonough.*

"What! do you really suppose that that animal had any feeling?" Thus Malebranche not merely believed he had not wounded or grieved her; he even thought he had caused her no physical pain. This was denying clear proof, and pushing faith in his master's doctrine to absurdity.

On the other hand, Montaigne, Leibnitz, La Fontaine, Bayle, Condillac, Madame de Sévigné, agreeing with all antiquity, from Pythagoras to Galen, assert that animals have all the organs of sensation and of feeling; that they possess will, desires, memory, ideas, combinations of ideas, and even the power of performing some moral acts, such as entertaining attachment like that a dog feels for his master, or a hen for her chicks; or, like "that very just equality which they practise in dividing food or other good things among their young," as Montaigne says; and that therefore the intelligence of animals, if not equal to man's, is at least like it, and that the differences between the oyster anchored to its rock and the *homo sapiens* of Linnæus are merely differences between more and less, degrees of succession that make up what is called the scale of being. It is the latter opinion that has been declared triumphant by the researches of natural history and those of comparative anatomy alike. On this point science has reached certainty, and every one, reading the story of "the two Rats, the Fox and the Egg," says now with La Fontaine:

"After that tale, where's the pretense
That animals are lacking sense?"

But another dispute has just been opened on this question: "Have animals the sense of beauty, the æsthetic sense?" The famous Charles Darwin, and his numerous followers agreeing with him, displaying even greater generosity than Montaigne, Leibnitz, and La Fontaine, answer unhesitatingly in the affirmative. In their view, just as animals are endowed with intelligence as well as man, though in a lower degree, so in the same way and in the same proportion they are endowed with the sense of beauty. They find the proof of this rather bold assertion, not in natural selection, which is a result of that struggle for life in which the weakest individuals among species, and the weakest species among genera, must disappear, but in sexual selection, which is a result of the struggle for reproduction, leading straight to the same consequences, namely, to the improvement of species and genera, which reaches, by slow elaboration through ages, even to their transformation. Among all animals, they say, among insects, fish, birds, mammals, the male chooses his female, and the female chooses her mate. If strength often determines the choice, so beauty often does too. The charm of graceful shapes, pleasing colors, fine notes, has great weight in settling the preference. Now, this charm which the male possesses habitually in a far higher degree than the female, he could have no occasion for whatever in the struggle for existence; he can benefit by it only in the struggle for reproduction. Hence all those displays of genuine coquetry which may easily be observed, in pairing-time, among all animals; hence those attractions prevailing through vigor of form, brilliant hues, and impassioned song. This general fact, well established under differences of appearance among all species and all genera, gives the Darwinian school ground for asserting that animals, having the perception of beauty, have consequently the æsthetic sense.

The opposing school, including of course all who prefer metaphysics to physiology, is by no means wanting in excellent answers. It says, by the pen of Charles Lévêque, "the doctrine of Darwin rests, in its last analysis, on the capital fact that the animal, the male sometimes, the female sometimes, often both at once, is susceptible to the beauty of its kind. That the animal is struck with it, we admit; but does it feel really this beauty of color, form, and song, in so far as it is beauty, or is it not rather the fact that those brilliant tints, that vigor, that sweetness of voice, form for the animal an indication, very expressive, yet merely brute-like, of a physiological condition which its instinct expects, which itself stimulates and answers to?" And this question, which must involve in its solution that of the former one too, the anti-Darwinians answer in the affirmative, which is the negative of the first question. They point out that this process of choice, in which beauty wins the prize, takes place only at the season of pairing, and ceases through all the rest of the year; that if we concede to the animal a sense of the beautiful, within those limits, still the sense has but one special object, finds its range within the species merely, never reaches to a less concrete conception, does not widen, does not perfect itself by tradition and culture, remains unchangeably set on the same object, the same time, the same point; that admiration which grows from the expansion of this sense of beauty can only spring from very varied and delicate comparisons; in one word, that the animal, being unable to attain general and abstract ideas, cannot rise to the æsthetic sense. To conclude: that sense is the exclusive endowment of man.

This is a concise statement—for neither our subject nor our powers allow us to take a side in this discussion of the interesting controversy raised by Mr. Darwin's last book. We only say that, notwithstanding the deep respect and the frank support that we give to his general doctrine, we here adopt the opinion of his opponents. We, too, believe that the æsthetic sense, in the high and perfect acceptation of the phrase, belongs to man alone. But we must here suggest a correction of the highest importance, which will at once define our opinion, and lead us to a fundamental reflection upon the arts.

"Complete the experiment," M. Charles Lévêque says. "Place your animal before a work of art representing its male or its female with a precision that deceives the eye; some of these works that seemed to live existed in the studios of ancient painters; they are more frequent in modern museums and exhibition-rooms. It was said that mares would neigh when going before horses painted by Apelles. A dog would perhaps stop a moment in front of Oudry's hunting-pieces, if their frames were put on the floor, within reach of his look. He would come up, examine them, ask the canvas a single question with his infallible scent—and that would be all. And yet, what is there in the picture? There is exactly the element worthy of admiration, that is to say, the expression of life by means of the most attractive colors and the most perfect forms. What does the quadruped care for these as he looks at this wonder? It is not the expression of life in general that he wants; it is life itself, individual life, life which speaks to his senses, and to that of smell much more strongly than to his eyes and ears. He has no concern with the general, the ideal, the admirable; he understands nothing about them."

This is all strictly and absolutely true. We have never put much faith in the stories invented by the Greeks and collected by Pliny; we have never remarked, in our long experience as a sportsman, that the most intelligent of man's companions ever gave the least attention to any object represented by painting or sculpture, even to his master's portrait, or that he looked with any feeling of satisfaction at one of those charming landscapes set before us at every step by Nature, enchanting as their view is to us. That seems to us one settled fact. But is M. Lévêque quite certain that *all* men have any greater faculty for that notice and that knowledge which are wanting in the dog? Have savages got above the level of animals in this respect? A traveler, who has studied Australia thoroughly, relates that at a gathering of some native inhabitants of that country he brought out portraits of Queen Victoria and Prince Albert, to see what they would think of them. Almost all kept silence, seeing nothing in the pictures at all within their range, or that touched their memory. But two of them, and the most intelligent ones among the tribe, such as can count as far as four, ventured on a remark. One said, "It is a boat"—the other, "It is a kangaroo."

All savages would answer in the same way, yet they are men. But, although they too love beauty after their fashion, though they have a strong liking for ornament, a great disposition for coquetry, savages do not go further than the special and individual object; they do not rise to a general, abstract idea; they know no more than the dog knows about seeing life in the representation of a living being. Therefore we need to correct the formula of Darwin's opponents, and may possibly thus succeed in bringing friends and enemies into agreement; their phrase must read, "The æsthetic sense belongs to civilized man alone.

In truth, it is not an innate faculty; it is a faculty acquired by tradition, by personal study, by the development of all the other faculties. The Australians must first learn to count as far as the number of fingers on the hand, and something more than that, before they can be able afterward to understand that pictures of a man and a woman, even if disfigured with a cap or crown, are neither boats nor kangaroos.

This assertion, that the æsthetic sense belongs only to civilized man, may be proved by both philosophic methods and by arguments which I take the liberty of stating concisely, as I have set them out at length in a treatise on the question "How should art be encouraged?" The subject is worth the risk of incurring ridicule for copying one's self. We may call it, if you choose, a second edition.

*A priori*, were we to assert that equality before the law means equality in intelligences, and that every man who has the right to be a citizen has the power to be an artist, we should commit a fundamental mistake. This would be confounding the feeling of the good in our nature with the feeling of the beautiful. The former is the instinctive knowledge of good and evil, of the just and the unjust; it is born with us, it is conscience itself, and, as such, it is necessary for all. The latter is a certain delicacy of sensation and of judgment which is formed very gradually in the course of life; it is called taste, and is useful only to a few. The feeling of the good, which marks the grand superiority of man over the animals, and forms the common basis of all societies, is an essential element in our nature, a gift we are forced to accept from the creative power. Without it, man would not be man. On the contrary, the feeling of the beautiful, which is less necessary, and may well be rare because it is superfluous, is an acquisition won by intelligence, slowly, painfully, uncertainly, and is often denied to the most honest efforts. One, like rank, costs nothing but the trouble of being born: the other demands, as all acquired knowledge does, a previous fitness, a kind of revelation, in which chance must often lend Nature its aid, besides time, reflection, mental labor in bodily leisure.

*A posteriori:* it is plain that a man's first movement, as regards the good, is almost always right; as regards the beautiful, almost always wrong. Listen to the crowd, judging by its moral estimate of an event that has taken place under its eyes, when no interest nor passion misleads it: what good sense, what fairness, what insight, what right intentions, what generous sympathies! Then hear it discoursing on the merit of works of art: what wretched taste, what glaring mistakes, what ridiculous enthusiasm, what sad and utter going astray!

"The people," writes Diderot to Grimm, "looks at everything, and gets at the meaning of nothing." But do you ask for still more decisive and clear practical proofs of the immense distance that separates knowledge of the good from that of the beautiful in the human mind? A petit jury, though it decides on the liberty, life, and honor of an accused person, is drawn merely by lot, because each citizen can judge as well as every other, upon evidence and discussion, of the truth and morality of an alleged fact. But do we choose the jury which has to award the prize of a competition from the same list and in the same way? No! in that case we must choose among the most skilled and the best qualified specialists.

But what need of demonstrating what his own experience tells everyone? Among these skilled and special jurors where is the one who will not confess that he at first was, and long continued to be, the dupe of his ignorance? Which of them is not aware that taste for the arts, and, still more, taste in the arts, came to him only after a long time, after lucky and often casual experiences, after protracted studies, repeated comparisons, a constant exertion of the powers of seeing, understanding, feeling, judging? And who does not know, by having learned it in himself, that in the arts—except perhaps in music, which filters in unconsciously to those who hear it—emotions come in the train of reasonings, and that the first condition of positive admiration is knowledge? "I am fully persuaded," says Sir Joshua Reynolds, in one of his "Fifteen Discourses," "that the pleasure yielded us by the perfections of art is a taste which we acquire only by long study and with much labor."

Those who choose to attribute too liberally to all men the sense of the beautiful as well as that of the good, attempt to support their opinion by a fact. They cite the instance of Athens, where, they say, competition in the arts was open to all in the public arena, where the whole people formed the tribunal. The instance is misleading, and I take it the other way, to sustain my proposition. Without insisting on the special genius of ancient Greece among the other nations of the world, and that of the Athenian people among the other peoples of Greece, I will merely point out that this people of Athens, so small in its territory and population, so great in its deeds and renown, consisted of about forty thousand free citizens, served by four hundred thousand slaves. Now, the slaves, charged with all manual tasks, and practising all the trades, relieved their masters from bodily labor, made them men of leisure, and consecrated them to the exclusive cultivation of the intellect, like the head of a body of which they were themselves the acting and subject limbs. That Athenian democracy, how jealous soever it might be of equality between citizens, was thus a real aristocracy; and we can very well understand how the decision on matters of art might be trusted to the multitude, when that multitude consisted entirely of men so enlightened, by education and experience, that all public functions and all magistracies might be distributed among them by lot, without any great danger to the state. The empire of the arts, though it refuses to accept any boundaries as haughtily as that of science does, though speaking as well a universal language it spreads over the whole world, can be ruled only by a strict oligarchy. Indeed, as has been justly said, good taste is the quintessence of good sense. Therefore, whatever may be our devotion to equality before the law, we can never admit equality before genius, or even before that talent which the cultivation of art requires.

Among mankind art has been, and always will be, the exclusive share of a small minority; in fact, a very choice and very limited aristocracy. It is assuredly not that of birth, for there is nothing more personal than genius or talent. It is aristocracy in its true and genuine sense—the privilege of the best. Art belongs only to certain exceptional natures, very rare because they are highly endowed, whose possessors must combine with the choicest moral faculties, imagination with judgment, feeling with taste, other precious physical faculties, clearness of view, and precision of touch. All men may be artisans; only the best, the *aristoi*, can be artists.

To conclude, and still taking the phrase "æsthetic sense" in its highest and perfect meaning, we first deny it to animals; and next, holding this delicate sense to be one of the noblest attributes of mankind, we restrict it to civilized man; then last, even among nations whom continued and general cultivation places in the front rank of humanity, we allow this aesthetic sense only to some groups of select men, on whom the nature of their minds and their taste formed by long study confers this rare and precious privilege. Has not Stendhal declared that all literary reputations are made by a choice circle of five hundred readers? And it is a certain truth that the number of bachelors of arts is still smaller than the number of bachelors of letters.

However, not to overstep the very modest part suited to mere writers and professional critics, let us hasten to add that, if in art the public voice is that of a very small number, but acute, practised, and disinterestedly enthusiastic; if this voice alone decrees to the living the rewards of celebrity, to the dead the immortality of fame; if, moreover, this choice circle of connoisseurs has the sole right of criticism, it by no means follows thence that every one of its members has the gift of infallibility. Far from that; it is like that aristocratic democracy of Athens, in which each citizen had his personal vote only, and could prevail only on condition of convincing. As in the domain of the good there is but one authority conscience, so there is but one authority—in the domain of the beautiful—taste. Only, conscience speaks the same language to all the men of one community; while taste, on the contrary, even acquired and formed taste, is as manifold as temperaments, ideas, and passions are. It varies from country to country in every age, and from age to age in every country. Still further, it varies between man and man, and in each man, at different periods of life. The bear in the fable very sensibly says:

"Who tells you this shape's awkward, that one fine?

Has yours the right to judge or censure mine?"

Therefore it is in vain, as if in strict pursuance of a duty, to read all books, listen to all counsels, ask advice from people supposed to be cleverer than one's self, and prop up one's judgment by more sure and authoritative judgments. In the criticism of art, where positive canons are wanting, no one is suffered to deem himself an authority; no one is anything more than an opinion.

# Part II

# The Aesthetic Sense and Religious Sentiment in Animals[2]

Dr Wilks reduces the chief difference between man and brute to the "smallness of knowledge of the fine arts possessed by the latter"; and a passing remark made by Prof. Huxley, in one of his essays, would seem to imply a disposition to draw the line of separation between animal and human intelligence at this point. Prantl regards the phrase "die Kunsttriebe der Thiere" as a metaphorical expression involving a confusion of terms, since animals, with all their apparent artistic ability and taste shown in constructing and decorating their

---

[2] By Edward Payson Evans.

habitations, do not seek to embody ideas in material forms—an assumption which begs the very question in dispute. Schiller, in his well-known poem, Die Künstler, makes man's pre-eminence consist solely in his artistic faculty:

"In Fleiss kann dich die Biene meistern,
In der Geschicklichkeit ein Wurm dien Lehrer sein,
Dein Wissen theilest du mit vorgezogenen Geistern,
Die Kunst, o Mensch, hast du allein."

In diligence the bee can master thee,
In skillfulness a worm thy teacher be,
Knowledge thou dost with higher spirits own,
But art, O man, thou dost possess alone.

Herbart, however, does not recognize this demarcation. "If one asks for a specific characteristic of mankind, which is not physical, but spiritual, original, and universal, and does not resolve itself into a more or less, I confess," he says, "that I do not know of any such distinction and do not think it exists." He then enumerates the advantages possessed by man—namely, hands, speech, and a long and helpless infancy, to the use and influence of which are due the extraordinary growth of the human brain in size and complexity and the corresponding development of intellectual power. In the acuteness of his senses and in many peculiarities of physical structure man is inferior to some of the lower animals. He has not, says Prof. Cope, kept pace with other mammals in the development of his teeth, which are "thoroughly primitive"; his nose is less serviceable than that of the dog; the eagle has a far better eye; the ankle joint of the sheep is, as a piece of mechanism, stronger and less liable to derangement than the corresponding joint in man; the horse's foot consists of a single compact elastic toe, on which the animal runs while its heel is carried in the air and never touches the ground, thus attaining a springiness and swiftness of motion beyond the reach of the human plantigrade. Whatever lightness and elasticity of step man possesses is due less to the perfection of his bodily organism than to the uplifting influence of his intellect. With the decay of his mental powers *Homo sapiens* slouches like a bear, as may be observed in the ungainly and unsteady gait of cretins and idiots, however vigorous they may be physically.

The objection urged by Prof. Kedny against the doctrine of evolution—namely, that man's helpless infancy proves him to be different in kind from other animals—ignores the fact that the soko and many other species of the genus *simia* pass through a period of infant helplessness almost as long as that of some savage tribes. The babyhood of the anthropoid apes is much longer and more helpless than that of the cynopithecoids, the platyrhines, or the lemurs; and the higher the order of the monkeys, the more they resemble man in this respect. Mr. Wallace captured a young orang-outang, which had to be fed and cared for like a human infant, lay rolling on the ground with all fours in the air, and could hardly walk when it was three months old; whereas a macacus of the same age seemed to have already acquired full use of its limbs and mental faculties. The long duration of this complete dependence on parental care in the case of the human infant, so far from disproving the doctrine of evolution, furnishes one of the strongest arguments in its favor, since it helps to explain how man gradually attained his intellectual primacy among the primates. The American platyrhines, marmosets, and other smaller long-tailed monkeys reach maturity in three or four years, whereas the African dog-headed apes require ten or twelve years for their full development, and with the larger anthropoids this period of growth is nearly as long as with human beings.

The fact that quadrumana have flexible organs of prehension, can grasp and handle things and imitate human actions, gives them a great advantage over quadrupeds. A dog may be as intelligent as a chimpanzee, but he is unable to "show off" as well; he can not untie knots with his paws, nor put on clothes, nor eat with knife and fork, nor uncork bottles, nor drink wine by lifting the glass to his lips, nor use a toothpick, nor perform a variety of tricks which make the monkey appear to be relatively far more richly endowed with mental gifts than is actually the case, and throw into the shade the most conspicuous exploits of the poodle and the collie.

Nevertheless, this manual and digital dexterity can scarcely be overestimated as a means of disciplining the mind and increasing the volume of the brain; and if chimpanzees, orang-outangs, and sokos had enjoyed the thousands of years of domestication and thorough breeding and training, from which dogs have so immensely profited, there is no knowing what advances in knowledge and acquisitions of intellectual culture they might not have made. It is wonderful how much they learn through observation and very slight instruction during a few months' intercourse with human beings, discharging with evident pleasure the duties of body servant or waiter, answering the door bell, showing visitors into the parlor, fetching water, kindling the fire, washing dishes, turning the spit, and doing all sorts of chores in and about the house. "Such an ape," said Brehm, "one cannot treat as a beast, but must associate with as a man. Notwithstanding all the peculiarities it exhibits, it reveals in its nature and conduct so very much that is human, that one quite forgets the animal. Its body is that of a brute, but its intelligence is almost on a level with that of a common boor. It is absurd to attribute the actions of such a creature to unthinking imitation; it imitates, to be sure, but as a child imitates an adult, with understanding and judgment."

That the plastic and progressive period of the monkey's individual development is short, and that its faculties become set and stationary at a comparatively early age, is undeniable; but the same holds true of the negro, who loses his educability and ceases his mental growth much earlier than the Caucasian. The longer or shorter duration of this formative season in the mental life of man is, to some extent, a matter of race, but in a still greater degree the resultant of civilization.

The hand is also a valuable instrument for the cultivation of the æsthetic sense, and the more flexible and sensitive this instrument becomes, the greater are the results achieved by it in this direction. But there are animals without hands that show an appreciation of the beautiful. Mr. Darwin has proved conclusively that birds take pleasure in sweet sounds and in brilliant colors, and that the sentiment thus awakened and appealed to plays an important part in the preservation and perfection of the species through natural selection. The struggle for existence is not always carried on by fierce combat and the triumph of brute force, but quite as frequently takes the form of competition in beauty, addressing itself either to the ear as alluring song or to the eye as attractive plumage; and the bird that possesses these characteristics in the highest degree carries off the prize in the tournament of love, and propagates its kind.

There is no doubt that birds take delight in the gorgeousness of their own feathers, and the more brilliant their hues the greater the vanity they display. Conspicuous examples of this love of admiration and fondness of parading their finery are the peacock and the bird of paradise.

The decoration of its boudoir by the bower bird, as described, by Mr. Gould in his History of the Birds of New South Wales, indicates a decided and discriminative preference for bright and variegated objects, and evinces no small amount of æsthetic feeling and artistic taste in selecting and arranging them. The bower is built of sticks and slender twigs gracefully interwoven, so that the tapering points meet at the top, and adorned with the rose-colored tail feathers of the inca cockatoo and the gay plumes of other parrots, tinted shells, bleached bones, rags of divers hues, and whatever gaudy or glittering trinkets may please the bird's fancy. Sometimes the space in front of the bower is covered with half a bushel of things of this sort, laid out like a parterre with winding walks, in which the happy possessor of the garnered treasures struts about with the pride and pleasure of a connoisseur in a gallery of paintings, or a bibliophile who has his shelves filled with *incunabula* and other rare editions. These objects have often been brought from a great distance, and are of no possible use to the bird except as they gratify its love of the beautiful and appeal to what we call in man the æsthetic sense. Its conduct can be explained in no other way; for the bower is not a nest in which eggs are laid and hatched and young ones reared; it is a *salon* or place of social entertainment, and thus serves a distinctly ideal purpose.

The singing of birds, as a means of sexual attraction, implies a certain appreciation of melody. Indeed, many of them do not confine themselves to the songs of their species, but learn notes from other birds and snatches of tunes from musical instruments. Canaries can be taught a variety of airs by playing them repeatedly on a piano or on a hurdy-gurdy. They listen with attention and imitate the strains which take their fancy. If harmony or the concord of sweet sounds, as distinguished from melody or the simple succession of sweet sounds, does not enter into bird music, the same may be said of the music of primitive man and of all early nations. Savages, like feathered songsters, sing in unison, but not in accord.

Not only do some species of monkeys, like the chimpanzees and sokos, get up concerts of their own in the depths of the forest, but dogs, which are generally supposed to be decidedly unmusical, also discriminate between tunes and express their preferences or aversions in an unmistakable manner. A friend of mine, who had a magnificent St. Bernard dog, was fond of playing the violoncello. The dog used to lie quietly in the room with closed eyes, and appeared to pay no attention to the music until his master struck up a certain tune, when the dog immediately and invariably sat up on his haunches and began to howl. If the tune which called forth such emotions had been written on a very high key, or characterized by shrill tones or harsh dissonances, the conduct of the dog might be easily explained. But such was not the case. There was nothing in this piece more than in any other, so far as anyone could observe, that ought to grate the canine ear. Many incidents of this kind might be cited to prove that even dogs are not indifferent to musical compositions, and show a nice discrimination between them, having their likes and dislikes, as well as human beings.

The fertilization and propagation of many plants depend upon the existence of a sense of color in insects, and the exercise of choice in the selection of flowers. This preference implies a pleasure in certain hues, and consequently the possession of a rudimentary perception of beauty. Plants whose fecundation depends upon the action of the wind do not develop such a variety of colors as those in which this depends upon the agency of insects. Nature can trust her ill-favored daughters to the wooing of the wind, but if she wishes to attract a nicer class of suitors she must endow her children with brilliant qualities.

The power of distinguishing between colors has been denied not only to the lower animals, but also to the lower races of mankind. But a more extended and accurate knowledge shows that the conclusion is incorrect in both cases. We know that the American aborigines discriminate between the seven primary colors, and it is absurd to infer that this faculty was wanting to the Homeric men merely because we do not find all these colors mentioned in the

Homeric poems. It has also been asserted that the ancient Assyrians could not distinguish green from blue or yellow, because no word was found for it in the remains of their language. But the tiles discovered at Nineveh prove that they had a very clear conception and æsthetic appreciation of the distinction between yellow, green, and blue, and probably did not confound any colors of the solar spectrum. The evidence of language on this point is purely negative and necessarily defective.

Even the religious sentiment, which has been assumed to be the peculiar possession of man, is faintly foreshadowed in the lower animals. The unanimity of opinion among those who have made the most careful study of this subject, and whose views are therefore entitled to the greatest consideration, is quite remarkable. M. A. de Quatrefages, in his *Rapport sur le Progrès de l'Anthropologie*, maintains that "domestic animals are religious, since they readily obey those who appeal to them with the rod or with sugar." In other words, they are amenable to rewards and punishments, doing the will and seeking to win the favor of superior beings, on whom they are dependent, propitiating and fawning upon them, creeping and groveling on the ground in abject adoration, in order to assuage their anger or to secure their kind regard. "There is no difference," adds the same author, "between the negro who worships a dangerous animal, and the dog who crouches at his master's feet to obtain pardon for a fault. . . . Animals fly to man for protection as a believer does to his god."

This is precisely the feeling of the savage in respect to the superior skill and power of the civilized man. *Taguta kipini te Atua*—doctor all the same as God—are the words in which the Morioris, or aborigines of the Chatham Islands, expressed their sense of dependence on a higher agency, whose beneficent workings they perceived but could not comprehend. Among rude tribes the sentiment of devotion to a chief does not differ essentially from that of devotion to a god; the Romans, at the height of their civilization, paid divine honors to their emperors; and in modern monarchies kings are officially addressed in terms of reverential awe and superlative adulation as all-wise and all-powerful beings, whose favor one can not sufficiently implore with servile words and suppliant knee.

"The feeling of religious devotion," says Darwin, "is a highly complex one, consisting of love, complete submission to an exalted and mysterious superior, a strong sense of dependence, fear, reverence, gratitude, hope for the future, and perhaps other elements. No being could experience so complex an emotion until advanced in his intellectual and moral faculties to at least a moderately high level. Nevertheless, we see some distinct approach to this state of mind in the deep love of a dog for his master, associated with complete submission, some fear, and perhaps other feelings."

Comte held that the higher animals are capable of forming fetichistic conceptions, and of being strongly influenced by them.

Herbert Spencer denies the truth of this statement in its absolute form, because it does not fit into his theory of the origin and evolution of religious ideas, but admits, what is essentially the same thing so far as the present discussion is concerned, that "the behavior of intelligent animals elucidates the genesis" of fetichism, and gives two illustrations of it. "One of these actions was that of a formidable beast, half mastiff, half bloodhound, belonging to friends of mine. While playing with a walking-stick, which had been given to him and which he had seized by the lower end, it happened that in his gambols he thrust the handle against the ground, the result being that the end he had in his mouth was forced against his palate. Giving a yelp, he dropped the stick, rushed to some distance from it, and betrayed a consternation which was particularly laughable in so large and ferocious-looking a creature. Only after cautious approaches and much hesitation was he induced again to lay hold of the stick. This behavior showed very clearly that the stick, while displaying none but the properties he was familiar with, was not regarded by him as an active agent, but that when it suddenly inflicted a pain in a way never before experienced from an inanimate object, he was led for the moment to class it with animate objects, and to regard it as capable of again doing him injury. Similarly, in the mind of the primitive man, knowing scarcely more of natural causation than a dog, the anomalous behavior of an object previously classed as inanimate suggests animation. The idea of voluntary action is made nascent, and there arises a tendency to regard the object with alarm, lest it should act in some other unexpected and perhaps mischievous way. The vague notion of animation thus aroused will obviously become a more definite notion as fast as the development of the ghost theory furnishes a specific agency to which the anomalous behavior can be ascribed."

This conduct of the dog, which everyone must have observed under similar circumstances, corresponds to that of the savage who worshiped an anchor which had been cast ashore, and on which he had hurt himself when he first came in contact with it. Superstitious fear of this sort prevails most among men of the lowest order of intelligence, or in that stage of society in which human beings are psychically least removed from beasts. In proportion as they rise in the scale of existence and unfold their mental faculties, the more they free themselves from the tyranny of the supernatural. The terror of the dog hurt by the stick was out of all proportion to the pain inflicted, and arose solely from the fact that it was produced by a mysterious cause; it was fear intensified by the intervention of a ghostly element, and thus working upon the imagination it assumed the nature of religious awe. The case is analogous to that of a big, burly, brutal savage trembling before a rude stock or stone, or a Neapolitan bandit cowering before an image of the Virgin or kissing devoutly the feet of a crucifix.

The other illustration given by Herbert Spencer is that of a retriever, who, associating the fetching of game with the pleasure of the person to whom she brought it, would often fetch various objects and lay them at her master's feet; and "this had become in her mind an act of propitiation."

Still more interesting and instructive are Mr. Romanes's experiments with a Skye terrier. This dog, which was exceedingly intelligent and therefore an excellent subject for psychological study, "used to play with dry bones, by tossing them in the air, throwing them to a distance, and generally giving them the appearance of animation, in order to give himself the ideal pleasure of worrying them. On one occasion, therefore, I tied a long and fine thread to a dry bone and gave him the latter to play with. After he had tossed it about for a short time I took the opportunity, when it had fallen at a distance from him and while he was following it up, of gently drawing it away from him by means of the long, invisible thread. Instantly his whole demeanor changed. The bone, which he had previously pretended to be alive, began to look as if it were really alive, and his astonishment knew no bounds. He first approached it with nervous caution, but, as the slow receding motion continued and he became quite certain that the movement could not be accounted for by any residuum of force which he had himself communicated, his astonishment developed into dread, and he ran to conceal himself under some articles of furniture, there to behold at a distance the 'uncanny' spectacle of a dry bone coming to life." In this instance we have the exercise of close observation, judgment, reason, and imagination culminating in the exhibition of superstitious fear—all the elements, in short, which constitute religious sentiment in its crudest form.

Animals are afraid of darkness for the same reason that children are. Thunder, lightning, and other violent meteorological phenomena, which inspire the primitive man with awe and therefore play a prominent part in the evolution of early mythology, produce a similar impression upon many of the lower animals, simply because they are mysterious noises which appeal to the imagination and stimulate the mythopœic faculty. Mr. Romanes states that "on one occasion, when a number of apples were being shot out of bags upon the wooden floor of an apple-room, the sound in the house as each bag was shot closely resembled that of distant thunder." A setter was greatly alarmed at the noise until he was taken to the apple-room and shown the cause of it, after which "his dread entirely left him, and on again returning to the house he listened to the rumbling with all cheerfulness." Dogs and horses can be completely cured of their fear of thunder by being present at artillery practice; they imagine that they now know what produces the dreadful roar, and are henceforth free from all apprehension concerning it.

To some extent this sense of the supernatural seems to enter into the sphere of pure imagination and to excite in the minds of animals those vague feelings of anxiety and alarm arising from mere figments of the brain and characterized as superstition. The following incident, "illustrating the instinctive fear of death and consciousness of its presence manifested by birds," is related by Buist: "A hen canary died, was buried, the nesting establisliment broken up, the surviving cock bird removed to a new cage, and the hatching cage itself thoroughly cleansed and purified, and put aside till the following spring. Never, however, could any bird afterward endure being placed in that cage. They fought and struggled to get out, and, if all in vain their efforts, they moped, huddling close together, thoroughly unhappy, refusing to be comforted by any amount of sunshine, companionship, or dainty food." The experiment was tried with foreign birds, that had not been in the house when the death of the hen occurred, and could not, therefore, have known anything of the melancholy event by observation. The result, however, was always the same. "For the future that cage to them was haunted."

It is a common belief that many animals can see ghosts and future events. Justinus Kerner declares (Die Seherin von Prevorst, i, 125) that they are endowed with second sight, and that numerous facts can be adduced in proof of it. This uncanny faculty is supposed to be especially strong in dogs and horses. Storks, too, are known to have foreseen the burning of houses on which they had been wont to build their nests, and to have abandoned them, taking up their abode on other buildings or on trees in the vicinity. No sooner had the anticipated conflagration taken place, and a new house been erected on the same site, than they returned and built their nests on it as heretofore. That Balaam's ass perceived the angel, which was beyond the ken of the prophet, ought to suffice to convince every believer in the plenary inspiration of the Bible of the specter-seeing powers of the lower animals. The ghost stories told of dogs and horses are quite as numerous and well authenticated as those which have been told of men. There is no psychological theory of apparitions that does not explain these strange phenomena as satisfactorily in beasts as in human beings. The night side of Nature casts its gloom over both.

Of course, if religion is a direct and special revelation to man, then no sentient creature prior and inferior to him could have any share in it. The hypothesis of a pure primitive monotheism, of which all polytheistic systems of belief are mere distortions and degradations, would also tend to exclude the lower animals from the possession of religious sentiment by showing that the religious history of the race has been a downward instead of an upward movement, a corruption instead of an evolution. Its growth would not correspond to the growth of intelligence, and it could no longer be studied as a psychological phenomenon, but would be removed at once from the province of scientific investigation. There can be no science of the supernatural, since science recognizes only the operation of natural laws. A miracle that can be explained, as the rationalistic school of theology has attempted to do, ceases thereby to be a miracle. The essence of religion is mystery; the sole aim of science is to clear up and thus do away with mysteries—a goal which it is always tending toward but will never reach, for the same reason that an asymptotic line never meets the curve which it is constantly approaching.

# Part III

## Aesthetic Feeling in Birds[3]

There is no portion of Mr. Darwin's great superstructure which has been subjected to more searching criticism than his theory of sexual selection—the theory that beauty in animals is dependent, in part at least, upon the choice of brightly colored, ornamented, or musically endowed mates by one or other sex among all the more highly developed classes, such as insects, crustaceans, birds, and mammals. Not only have opponents argued strongly against the existence of such æsthetic tastes in the lower animals as would account for the supposed preference

[3] By Charles G. Allen.

for beautiful partners, but even many of those who accept the evolutionist hypothesis as a whole have declared themselves unable to give in their adhesion to this particular speculation. Professor Mivart has brought forward strong objections to the great naturalist's view, and Mr. A. R. Wallace has raised a counter-theory on the subject of coloration at least, which has done much to convince many wavering biologists, and to insure their rejection of the suggested cause as adequate for the production of that beauty which all alike recognize in the animal world. It seems to me, however, that a little too much stress has been laid upon the notion that comparatively advanced intelligence is necessary for the appreciation of beauty in the opposite sex. It is true that our own highly complex æsthetic feelings are largely composed of elevated intellectual and emotional elements; but it may perhaps be shown that aesthetic tastes quite sufficient for the production of the known results do actually exist in many cases, and consist almost entirely of very simple sensuous factors. In order for a butterfly or a humming-bird to admire its gorgeously appareled mate, it is not necessary that it should be capable of taking delight, like ourselves, in a Claude or a Rubens; it is enough that it should possess a nervous organization pleasurably affected by certain forms and colors in the same way as it is pleasurably affected by sweet fruits or the nectar of flowers. Nothing more than this need be postulated in order to establish the facts for which Mr. Darwin has contended with such wealth of illustration in the second part of the "Descent of Man."

It may be worthwhile, then, to examine a single large class of animals, in which the æsthetic nature is highly developed, for the purpose of discovering whether they do really afford proof of a sensibility to form, color, or musical sound. It must be remembered that even in our own race the sense of beauty in children, savages, and the uncultured classes, hardly rises above this simple level. We must not, of course, expect to find an appreciation of musical harmony, of imitative pictorial skill, of elaborate ornamentation, among birds or insects. We must be content if we see evidence of a love for red, blue, and yellow, for sweet perfumes and pleasant flavors, for symmetrical forms and simple patterns, for ringing notes and trilled resonances. The class of birds probably shows external marks of such tastes in a higher degree than any other; and, though many of them have been set forth by various writers elsewhere, it will perhaps repay the trouble to collect them into a single paper in order to show their bearing upon the general æsthetic sensibility of the class, as well as upon the specific question of sexual selection. For this purpose I shall first take for granted the fact of such selection, and afterward endeavor to justify it by analogy from known human practice.

Beginning with the lowest of the special senses, taste, we find ample evidence that very many birds have a strong liking for sugar. In confinement, canaries and parrots eagerly devour it in the manufactured form.

In the wild state humming-birds, sun-birds, honey suckers, lories, and many other species, feed off the nectar of flowers, more or less mixed with insects. Mr. Webber, an American naturalist, found that the ruby-throats of the United States were attracted by a cup of sirup, and numerous other birds display a strong liking for the same mixture. Fruits, which have been developed especially to suit the tastes of birds, almost always contain an abundance of sugary juices; while the kernels within their stones are generally bitter, so as to prevent their winged allies from devouring the actual seed. Hence we may infer that all the vast tribes of toucans, hornbills, macaws, plantain-eaters, birds-of-paradise, and fruit-pigeons, possess a taste for sugar sufficiently strong to have produced the separate evolution of these sweet seed-coverings in a hundred different families of plants throughout the whole world. Indeed, the strength of the evidence thus afforded cannot be overrated, when we remember that in every case the covering is a dead loss to the plant, except in so far as it aids the dispersion of seeds; and that it must have been developed over and over again in a thousand different cases by the action of the most widely different birds. It is impossible to believe that such a coincidence can be due to accident, impossible to doubt that it results from a genuine taste for sweet flavors.

There is even some reason to believe that birds care for and discriminate other tastes besides the fundamental distinctions of sweet and bitter. All the small birds in

Jamaica are particularly fond of the little scarlet capsicums grown in gardens, and devour them so greedily, that the fruit has acquired the common name of *bird-peppers*. If we remember how very hard is often the almost horny covering of a bird's tongue, there is nothing remarkable in the fact that the pungency of the capsicum should be felt as an agreeable stimulant, probably having effects analogous to those of mustard, water-cress, or peppermint, with human beings. The oft-quoted liking of tropical pigeons for the nutmeg, with its aromatic coating of mace, points in the same direction. Parrots in captivity frequently display very decided preferences and antipathies in their food. Owls cannot be induced to taste meat in the slightest degree tainted. Again, all birds have a most accurate notion of the difference between ripe fruits and the unripe sour ones, besides carefully choosing the sunny side of peaches, pears, and apricots. The very frequency of distinct sapid principles in fruits would seem to favor the same supposition, as they have probably been acquired for the special allurement of particular species. Indeed, the more we consider the origin and nature of succulent fruits, the more does it become clear that they have been developed to suit the tastes of animals having essentially identical sensations with our own.

The case of the nutmeg leads us naturally on to the consideration of smell. Here we may conclude with great probability that the large class of aromatic fruits has acquired its perfume for the sake of attracting birds, especially when we recollect that flowers have acquired exactly similar perfumes for the sake of attracting insects. And although the possession of scent as a means of sexual allurement is rare among birds, being probably confined to the musk-duck and a few other species, yet it occurs frequently among butterflies, and is represented among mammals by the musk-deer, beaver, and many other ruminants or rodents. Curiously enough, the similarity of taste thus testified extends to the vegetal world in the case of the musk-plant; while even certain carnivores, such as the cat tribe, are extremely fond of "valerian, lemon-thyme, camomile, lavender, and many plants rich in essential oils." On the other hand, a good observer notes that cats have their dislikes, and he has often seen a tabby "smell at a fig-tree, and turn away with the disgusted air of a connoisseur." We have no such strong facts in the case of birds, but the frequency of perfumes in those fruits which depend upon them for the dispersion of their seeds, coupled with their total absence among most nuts, would lead to the conclusion that their likes and dislikes in the matter of smell are fully as marked.

But it is when we arrive at the sense of hearing that we come to the point where proper æsthetic feelings begin. It is quite impossible to doubt that birds are fond of musical sounds. The song of our own nightingales and linnets, the deep notes of the South American bellbird, the incessant cooing of the dove, the noisy chattering of the parrots, the ringing cry of the whippoorwill, all lead to the same conclusion. Here, again, these sounds are of precisely the same nature as those employed by the crickets, katydids, cicadas, and other musical insects, as well as by man himself in his vocal and instrumental music. Something of the same taste is displayed among the quadrumana by the howlers and other monkeys. But it is a noteworthy fact that a large majority of these presumably sexual calls, in birds, insects, and other animals, are true musical sounds, not mere noises. I have pointed out elsewhere the probable reason for this preference of pure tones in the case of mankind; and the same argument will apply, *mutatis mutandis,* to all other animals. But there is certainly a singular analogy in this respect between sounds and colors, most animals preferring the relatively pure and simple musical tones to confused noises; and the relatively pure and simple analytic colors, red, blue, green, and yellow, to confused mixtures such as brown, gray, and mud-color. At any rate, a bird evidently pays far more attention to the musical class of auditory perceptions than to mere noise. A canary will take no notice of ordinary confused sounds in a room; but, if one begins to chirp or whistle to it, it immediately responds with another chirp in emulation. So, too, when a piano or other musical instrument is played in the neighborhood of a singing bird, it will often show its recognition of the musical character by pouring out its very fullest flood of song, as if to conquer its unconscious rival. Of course, the singing-matches between birds themselves are too familiarly known to call for separate mention. It may be worthwhile, however, to notice that this love of musical sound exists even among certain reptiles; for I have often seen the common house-lizard of Jamaica listening with evident interest and attention to the playing of a piano, turning his head from side to side, and scampering away when disturbed, only to return again to the fascinating sound after a minute or two of hesitation.

The cases of the starling, the piping bullfinch, and the mockingbird, which can be taught to whistle a tune, show the same power still more highly developed. These instances prove not merely susceptibility to musical sounds, but also a capacity for distinguishing the harmonic intervals. It is stated that some birds, even in the wild state, display considerable knowledge of the musical scale; and a San Francisco naturalist is at present engaged upon a work in which he hopes to show that the human ear possesses in this respect merely a more highly developed form of the common vertebrate sensibility. When we reflect upon the purely physical and physiological basis which, as Helmholtz has taught us, underlies the musical intervals and the distinctions of harmony and discord, there is certainly no reason why they should not be perceived by all the higher animals alike, in a greater or less degree.

Considering, therefore, the evident susceptibility of birds to the simpler pleasures of music, and the interest which they show in it even apart from their domestic relations, there is no *a priori* difficulty in accepting the belief that their powers of song may have been developed by mutual selection, provided no adverse argument can be shown against the probability of such selection ever proving a cause of specific variation. To this last question, the question so ably raised by Mr. Wallace, I shall return on a later page.

Passing on to sight, we have first to observe the effects of mere light or brilliancy upon birds, apart from special effects of color or form. Now, birds certainly share with insects and many other creatures the common fascination for bright lights. "Owls and night-jars have been known to flutter against the window of a lighted room in the small hours." In the tropics, where windows are more constantly left open, birds frequently fly into houses, attracted by a lamp or candle. The reflected light of a mirror is employed to draw down larks. Magpies delight in secreting diamonds, gold, silver, and other shiny objects. The bower-birds use shells, polished pebbles, and like brilliant odds and ends in the construction of their bowers. So, too, metallic iridescence occurs frequently in the feathers of beautiful species, notably in the humming-birds, sun-birds, peacocks, and other flower-feeding or fruit-eating classes. But even the far less brilliant crows, gulls, ducks, and doves show exquisitely burnished gloss or luster on their coats, often specialized upon particular portions of the plumage, and apparently betraying the action of sexual selection.

Of the love for color shown by birds, I have already treated so fully elsewhere, that it will suffice here briefly to recapitulate the main facts. The universality of bright hues in the fruits which depend upon birds for the dispersion of their seeds clearly shows that fruit eating species are attracted by red, blue, purple, and yellow; just as the analogous case of insect-fertilized flowers shows the preference of bees and butterflies for similar tints. Mr. Darwin has collected several instances of interest displayed by birds in colored objects-, and of the attractiveness which color evidently possesses in their eyes. Of these, the most remarkable cases are those of the bower-birds' and the humming-birds' nests. And the constant occurrence of very brilliant hues among flower-feeding species, such as humming-birds, sun-birds, lories, and barbets, or among fruit-eaters, such as toucans, fruit-pigeons, birds-of-paradise, and parrots, induces the belief that in these classes the exercise of the structures upon the search for food has led to the formation of a very strong taste for color, ultimately resulting in sexual modifications.

As for the harmony of color usually observable in birds, it must be remembered that our feeling of harmony probably depends upon the due intermission and alternation of sense-stimulants, and therefore ought naturally to be shared by us more or less definitely with all other animals having a like constitution of the eye. Now, the mammalian and avian eye being derived from a common ancestor, who already possessed a highly developed power of vision, we might reasonably expect that our feelings of harmony would be essentially identical; and this expectation is fully borne out both by the coloration of fruits and of birds themselves, which seldom or never present what we should regard as discordant coloring. Furthermore, as the most beautiful classes of birds are those which live perpetually among tropical flowers and fruits, in the most beautiful forests or meadows, surrounded by exquisite insects and reptiles, and forever exercising their vision upon the most diversely colored environment in the whole world, it would seem far from impossible that their chromatic sensibility is even more highly developed than that of average humanity, and therefore that harmony or discord of colour would bear a relatively greater importance in their eyes than in those of any human being except the most artistically endowed. This conclusion will doubtless sound strange and even grotesque to those who are always accustomed to postulate for man a kind of absolute supremacy in the scheme of Nature; but it appears to me almost as obvious and as simply accounted for as the superiority of scent in the dog and the deer, or of distant vision in the eagle and the vulture. Lastly, it may be noted that much of the beauty of birds, as of insects, fruits, and flowers, is due to the delicate gradation of tints which they display. But in all natural products such gradation is an almost necessary result of the mode by which they have been evolved. It is only in human manufactures, where pigment is laid on with a brush or stamp, that colors can be placed in crude juxtaposition to one another, giving rise to the worst form of chromatic discord.

Doubtless our native feeling of dislike to such discords, based upon their immediately fatiguing effect upon the nerves employed, has been heightened intellectually by the knowledge that they differ so widely from the dainty gradations to be found in the handiwork of Nature. Besides being sensuously recognized as discordant, they are intellectually recognized as inartistic. Thus a large part of our art-progress has consisted in an advance from the harsh and monotonous fields of primary red and blue, divided by very hard and definite lines, which we find in Egyptian painting, to the faithful representation of graduated tints and shades which appears upon a modern canvas. But in the petal of a rose, the ray of a daisy, the wing of a butterfly, the tail-covert of a peacock, such gradual merging of tint in tint could hardly fail to occur spontaneously, as a product of evolution; while the comparatively definite marking off of special spots or lines, as in some orchids and other flowers, could only present itself as a result of very intense competition between species, carried on under highly complex conditions. The views set forth by Mr. Bates upon the progressive modification of patches or regions in a butterfly's wing, and by Mr. Darwin and Mr. Wallace on the feathers of the peacock and the Argus-pheasant, though widely differing as to the particular mode of their evolution, yet alike convince us that the inevitable result must be just such a graceful running together of contiguous colors as we actually find to obtain in every case.

Lastly, we arrive at the sensibility to form, symmetry, arrangement of patterns, and the like higher sensuous aesthetic feelings, which remains in the eyes of many the chief stumbling-block in the way of accepting the theory of sexual selection. The pleasure derived from sweet tastes and fragrant perfumes is so purely sensuous that nobody doubts its universal existence among all the higher animals. The pleasure derived from musical sounds and bright colors, though more intimately bound up in the human consciousness with intellectual and higher emotional elements, yet contains so large a factor of mere sensuous stimulation that we can easily conceive of it as appealing to the ears and eyes of insects and vertebrates. But the still higher pleasure derived from graceful curves, symmetrical ornamentation, and elaborate tracery is so largely made up of intellectual feelings, and so largely supplemented in our own case by associations of costliness, human handicraft, or imitative skill, that we find it hard at first sight to believe in the existence of similar feelings among pheasants of the Indian jungle, antelopes of the African plains, or monkeys of the Brazilian forest. Even here, however, a little consideration may convince us that the æsthetic appreciation of form and its connected varieties is not necessarily above the narrow intellectual faculties of the higher vertebrates and articulates at least.

In the first place, if we look at the human race itself we shall find that a comparatively high susceptibility to form occurs even among very low races. Indeed, most exquisite patterns are produced by savages whose taste in color is apparently far less developed than that of parrots, humming-birds, and fruit-pigeons. The tattooed tracery of the Polynesians and many other savage tribes presents beautiful designs of which even a European decorative artist need not be ashamed. The New Zealand canoes, the paddles and clubs of the Admiralty-Islanders, the shields of the Zooloos, are all most graceful in their shapes and most daintily wrought with interlacing patterns in carved work. Calabashes, cocoanuts, ostrich-eggs, and other early vessels are always cut in sections which exactly coincide with the demands of the most developed taste. The huts of savages are generally square, circular, or oval in shape, neatly wattled at symmetrical distances. The earliest architecture consists of regular stone rings, avenues, tumuli, and other definitely shaped monuments. Dr. Schweinfurth's "Heart of Africa" contains pictures of pottery as beautiful as anything ever produced in Greece or Etruria, stools, chairs, and other furniture as gracefully shaped as anything ever wrought by a Renaissance carver, and villages as prettily arranged after their simple fashion as the architects of the Parthenon or Cologne could have arranged them. If we look back in time, we find the stone hatchets and arrow-heads, not only of the neolithic but even of the palæolithic age, carefully symmetrical in shape, and that at a time when the extra labor of chipping the flints into comeliness must have entailed a considerable waste of human or half-human energy. At the same early date we find fossil shells, symmetrical bones, teeth, and other like objects, already drilled to serve as necklaces or other ornaments, which analogy with the similar ornaments now in use would lead us to believe were symmetrically strung together into definite patterns. Indeed, the more we look at the products of the very lowest savages and the very earliest men, the more shall we be convinced that they possessed in the germ all those aesthetic feelings which have finally developed our existing architecture and other decorative or semi-decorative arts.

Again, we cannot fail to be struck by the fact that man has always employed for ornamental purposes exactly those very appendages of animals which, if the theory of sexual selection be correct, have been produced by the animals themselves as ornamental adjuncts. The feathers of peacocks, the plumes of the ostrich and the bird-of-paradise, the antlers of deers, the horns of antelopes, the tusks of elephants, mammoths, and musk-deer, the striped, spotted, or dappled skins of mammals, all these have been used from the earliest periods as materials for decoration by mankind. Exactly the same curls, twists, and patterns which seem to please the eyes of animals are known to please the eyes of man, even in his lowest developments. If these ornaments were not produced because the creatures themselves found them beautiful, at least they are the same as those which would have been produced had the taste of such creatures coincided in the main with that which runs throughout the whole of humanity, from the most degraded savage to the highest artist.

Moreover, part at least of the pleasure of form probably has a purely sensuous origin. The superiority of curved lines to straight, of the waving or sinuous contour to the angular, is apparently connected with the muscular process in the act of vision. Hence there is no reason why it might not be felt by intelligent animals, just as we know that it is felt, and acutely felt, by hardly more intelligent men.

Similar conclusions are forced upon us if we look at the nature of the supposed ornaments themselves. They are almost always, like the horns of several ruminants, the tail-coverts of the peacock, and the lappets or crests of many birds, apparently devoid of any functional use whatsoever, unless that use be the attraction of the opposite sex. They are also marked by the extreme definiteness of their shape, color, or sculpture—a definiteness which never occurs in similar structures among the lower animals. For though some echinodermata, as for example the sea-urchins, are very beautifully and regularly marked, yet their markings are purely dependent upon the structural arrangements of the animal, and cannot generally be detected till after death. So, too, the shells of many mollusca, such as scalaria and the murices, are very beautifully sculptured; but this sculpture is structurally necessary for the animal, and apparently depends entirely upon the shape and markings of the mantle. Among birds, however, as among the ruminants, all the structures ascribed by Mr. Darwin to sexual selection are marked by a kind of definiteness, quite unconnected with ordinary functions, which it is difficult to describe in words, but which can immediately be *felt* if we compare the coloration of a peacock with that of a sea-anemone or a medusa. The former is perfectly definite without being obviously connected with structure; the latter is very indefinite, and yet bears a clear relation to the general shape of the animal. This combination of great specific distinctness with little apparent functional value appears to me the genuine hall-mark of organs due to sexual selection.

There is even some little external evidence in favor of a love for symmetry among birds. The nests of weaver-birds and many other species, as well as the bowers of the bower-birds, display a considerable taste for orderly arrangement. For one must remember that the building of such nests, though doubtless instinctive and inherited, is not a mere organic process, like the secretion of a molluscan shell; it is as much an art as the building of a honeycomb or of a savage hut. The flight of birds in play, the antics of many humming-birds, the strange eddyings and aërial evolutions of several other species, all approach very nearly to our own idea of dancing. I am almost afraid to hazard the observation, yet on the other hand I cannot avoid risking it, that the attitudes taken up by the turkey-buzzards or John crows of the West Indies upon the tops of houses frequently seemed to me intentionally symmetrical. I have observed them sitting in every variety of position—

one at each end of a long roof; one at each of the two points half-way between ends and middle; three arranged in either of these forms, with one in the middle; five arranged in the order C, B, A, B, C, etc. If any other observer can supplement this experience, which I record with great diffidence, I shall be very glad.

Taking for granted, then, this appreciation of form and symmetry, we shall find that it has produced many notable effects in the world of birds. To it, apparently, we owe the crests of cockatoos, pigeons, herons, and a hundred other species; the wattles, combs, hackles, and lappets of the gallinaceous birds; the beaks of toucans, hornbills, and cassowaries; the wonderful marking of the peacock and the Argus pheasant. Anyone who wishes really to understand the immense variety of ornamentation which has thus resulted should pay a visit to the ornithological rooms in the British Museum, and observe the innumerable devices for attracting attention which exist in almost every order of birds. Perhaps the familiar lyre-bird offers the very finest example of all, so far as beauty of form and symmetry of arrangement are concerned. It is specially noticeable, however, that in almost every case the decorations are lavished on the very same parts on which they would have been bestowed by human taste.

If, then, we put together all the scattered indications thus afforded us, if we consider the taste for sweet food and delicate perfumes, the song of the nightingale and the graceful movements of the swan, the metallic colors of the flower-feeders, the exquisite hues of the fruit eaters, the varied plumage of the birds-of-paradise, the beautiful nests and bowers, the habit of abstracting brilliant objects, the universal loveliness of shape or tint throughout the whole class—we can hardly doubt that birds, as a whole, possess aesthetic endowments of a very high order. Let us proceed to consider the general bearings of these views upon the question of sexual selection.

Mr. Herbert Spencer, in a very remarkable essay upon personal beauty, has shown that in the human race we regard as beautiful, on the whole, just those personal peculiarities which are, roughly speaking, the external marks of fitness for the conditions of human life. More especially do we admire those points which bespeak a *physique* adapted for the duties of paternity and maternity. We dislike excessive leanness or excessive fat; a sallow or a bloated complexion; deformity or extreme departure from the normal type. On the other hand, we like in man robust and muscular limbs, an erect carriage, an open chest, a virile development of beard and whiskers, with all the other outward signs of health and strength. We like in woman a womanly and tender face, a fine and well-developed figure, and all the other outward signs of health, and more especially of healthy maternal capacities. We like in both sexes an abundant crop of hair, clear and bright eyes, white and well-set teeth, red lips, and cheeks which show a good and sound circulation; we like an expression which betokens good humor, moral qualities, and refinement; lastly, we like a face which indicates intellectual power and ability to succeed in the highly complex struggle for life in the midst of which our lot is cast. One or other of these points we may occasionally waive in consideration of other special claims; but if anybody asks in the abstract whether we prefer a stunted *physique* to well-grown limbs and muscles; a flat-chested woman to one with a finely-proportioned bust—unhealthy and sallow skin to a clear complexion; a sour-looking, mean, or brutal face to a bright, joyous, open, and honest countenance; silly or idiotic features to an expression full of liveliness and intelligence—there can be but one answer possible. Leaving out of consideration for the present all other elements of the involved and complex problem, we may conclude that beauty, from one point of view at least, consists for each species in the outward signs of specific adaptation to specific necessities.

On the other hand, beauty also consists from a different point of view of stimulation by a certain relatively fixed number of external stimulants—musical sound, brilliant light, analytic colors, curved shapes, symmetrical arrangements of form, etc.—which appear to act directly upon the nervous system. This is clearly the view which Mr. Darwin implicitly accepts, especially with regard to tone and color. The facts at which we have briefly glanced above respecting the aesthetic feelings in birds, and the beauty of the birds themselves, take for granted some such theory of the aesthetic faculty. How are we to find a reconciliation between this view and that of Mr. Herbert Spencer?

I believe the true clew has been given us by Mr. A. R. Wallace, in the able essays on "Color in Plants and Animals" which originally appeared in "Macmillan's Magazine," and were afterward reprinted in his work on "Tropical Nature." It is true that Mr. Wallace utterly rejects sexual selection as a *vera causa,* and substitutes for it several separate minor modifications of natural selection; yet it seems to me that a compromise between his view and the two other views of Mr. Darwin and Mr. Herbert Spencer would more really represent the actual state of the case in nature. Or, to put it more correctly, the three ideas are not in reality contradictory or even opposite, but are rather different and complementary aspects of one and the same fundamental truth.

Beauty in the abstract and for all species, as it seems to me, consists of pleasurable stimulation of the higher sense-organs. Such pleasurable stimulation must, on the average of cases, be given rather by brilliancy than by dullness; rather by analytic colors than by confused hues; rather by curved or flowing forms than by angularity; rather by musical sounds than by mere noises. But beauty relatively to the particular species, and especially as regards the sexual relation, must be largely due to special inherited tastes, doubtless ingrained and physically registered in the nervous system, leading the animal to derive pleasure from the typically healthy and normal form of the opposite sex. For, if any individual possesses divergent tastes, they must either be for relatively unhealthy and typically defective forms, in which case they will tend to be promptly suppressed by natural selection; or for neutral or improved forms, in which case they will help to give rise to new varieties, ultimately culminating in separate species. Such divergent tastes seem to be shown in all large dominant families, such as the humming-birds, where specific variation and ornamentation have been carried out to a very great extent. But all such divergent fancies must themselves tend to become distinctly fixed for purposes of specific identification; and we find as a matter of fact that each species does readily recognize its mates, even when the differences between closely allied species are only very slight.

Now, this special hereditary liking for a particular form and type will not interfere with the general love for color, brilliancy, sweet tones, and perfumes. Accordingly, wherever the circumstances which give rise to a taste for these sense-stimulants exist, it would naturally follow that the taste would help to determine the choice of mates. But, again, as Mr. Wallace has fully shown, the most vigorous individuals would usually possess the most highly developed ornaments, the brightest colors, the largest scent-glands, and the loudest or most musical voices. Hence the very animals most likely to be sexually selected are also, on the average, those most likely to be naturally selected. Yet sexual selection really differs from natural selection, in that it gives a special direction to the ornamentation. For example, one can hardly believe that mere masculine vigor will account for the gorgeous and positively inconvenient plumage of the bird-of-paradise, nor for the exquisite coloring of the peacock, nor for the extremely ungainly air-bladders of many insects. It is quite easy to conceive that the general vigor implied by the possession of these extended ornamental adjuncts may have helped their possessors in the general struggle for life; but it is hardly possible to believe that they could have reached their present definite development without the aid of sexual selection. In short, where an ornament, or what seemed to any particular individual an ornament, proved hurtful to the race, it would be eliminated by natural selection; but where it proved neutral it would be spared, and if it coincided with advantageous qualities it would be further developed. Yet, even if only neutral, sexual selection alone would give it an extra chance, and, as it would doubtless be correlated on the one hand with certain special tastes and habits, and on the other hand with certain slight modifications of structure, it would doubtless succeed on an average of cases in producing a new species.

The familiar facts of human beauty will probably serve to make this reconciliation of the conflicting views a little clearer. Man of course admires in the abstract bright colors, brilliancy, musical notes, graceful curves, and symmetrical form. But, as applied to the human face and figure, he admires these in certain special and typical arrangements. Thus, while our general love for color leads us to prize golden hair, we do not like a sallow complexion; while it leads us to see beauty in rosy cheeks and red lips, which are signs of a healthy circulation, we do not admire the same redness in the nose, where it is usually a result of dyspepsia or dissipated habits, either of which is bad for the race at large. Again, though we admire pearly teeth, clear eyes, and a white skin, all of which are obviously the external marks of useful properties, we do not admire white cheeks, which are the external mark of weakness or anæmia. Similarly, our idea of beauty demands that the figure should neither be too fat nor too thin, but should possess that graceful development of all the muscles which is the outward symbol of ability to move and act with ease and effect. If any large number of persons were ever actuated by opposite tastes, if they preferred pale cheeks and lips to rosy ones, thin and haggard faces to full and rounded ones, weak and angular limbs to strong and graceful ones, a flat and undeveloped chest to a fine and healthy bust, then they and their taste must rapidly die out through the inferior *physique* they would hand on to their descendants. And as every individual is himself the product of countless thousands of prior individuals, all of whom have been in the main successful in the struggle for life and the search for mates, it must follow that he will have inherited from them, on the average, a healthy taste for that particular arrangement of limbs and features which best suits the essential conditions of the species. Not, of course, that he will consciously recognize this fact in most cases; but the mere presentation of such a typical combination will instinctively rouse in him, through the organized correlation of nervous centers, the hereditary feeling of beauty. Hence this feeling will probably be most strongly aroused in each species by the sight of the sex which in that species has undergone the greatest differentiation through sexual selection: just as we know that the feeling is most strongly aroused in mankind by the beauty of woman. On the other hand, we are still able to perceive, when we look at a peacock or a humming-bird, that, thought his specific hereditary feeling is absent, yet the strength of the purely abstract elements—color, brilliancy, symmetry, form, and minute workmanship—is so unusually great that we have no hesitation in pronouncing them also beautiful after their kind.

If, then, we admit the reality and potency of sexual selection, in however modified a form, it must follow that birds, being on the whole the most ornamental of all classes in the animal world, are also the most æsthetic, with the exception of man. It might, at first sight, seem that consistency would demand the sacrifice even of this exception; but a moment's reflection will disclose an important difference between the two cases. Man possesses the active power of direct artistic creation; the birds only possess the passive power of selection from among the forms produced for them by Nature. The ordinary workman who selects his wife partly or wholly on the ground of beauty, thereby does something toward perpetuating and improving the beauty of the race; he stamps the impress of his taste upon future generations; but such mere passive choice differs widely from the ability to depict or create on canvas such a beautiful woman. In this way, the actual loveliness of birds may lead us somewhat to over-estimate their aesthetic sensibility; for, though within their own species they may be capable of distinguishing between comparatively minute shades and degrees of beauty, just as we can distinguish between such minute points in human faces as would doubtless absolutely escape the notice of any other animal, it is yet improbable that they would be equally discriminative outside the limits of their own species. Again the principle of "gradation of characters" necessitates certain artistic effects in their plumage which they themselves may be only half able to admire. So, too, the necessarily symmetrical arrangement of the two sides of the body and the mode of growth of feathers may often have helped, unintentionally, as it were, in producing the total effect. In other words, it may well be that the birds, while selecting their partners on the ground of bright color, exceptionally long plumes, and other ornamental characters which they *could* understand and admire, may have succeeded in producing harmonies of tone, delicate gradations of tint, and other similar effects which they could *not* understand or admire, or at least could only admire very partially.

Yet, after making all allowances for possible reading in of human feelings, it may probably be asserted with safety that the actual appearance of birds entitles them to rank, on the whole, higher in the aesthetic scale than any other animals except man. Whether we look at their graceful shapes, in the swan and the heron; their beautiful plumes, in the ostrich and the bird-of-paradise; their exquisite color, in the sunbird and the lory; their ornamental crests and lappets, in the humming-birds, the pigeons, and the parrots; or their song in the linnet, the mocking-bird, and the nightingale—we must confess that they give extraordinary evidence of a taste for all that man considers lovely or artistic. And this is just what we might expect from their free mode of life, their rapid motion, their highly developed senses, their comparative freedom from enemies, their long and almost uninterrupted rivalry between themselves for the possession of their mates. Especially should we expect this splendid outburst of aesthetic sensibility exactly where we find it in its greatest glory, among the flower haunting and fruit-eating species of the Brazilian forests, the Indian jungles, and the Malay Archipelago. Surrounded for generations and generations by gorgeous orchids and trumpet-creepers, from which they sucked the stored-up nectar, by gleaming purple or golden fruits, by burnished beetles, metallic butterflies, bronze-scaled lizards, and coral snakes, their prey or their enemies, exercising their eyes perpetually in the search for food among the exquisite objects of their environment, and safe from almost all foes except those of their own class, tropical birds have naturally developed the most gorgeous and the most perfect forms and colors in the whole animal creation. And, above all, they have stamped the mark of their peculiarly high aesthetic feelings upon their own shapes by the wonderful definiteness of their patterns and their ornamental adjuncts, nowhere equaled, save in the most perfect decorative handicraft of man himself.

Made in the USA
Monee, IL
08 November 2026

67378649R00073